Solution of Ancient Greek Famous Construction Problems in classical Geometry

Unique works on Euclidean geometry

By
Manoranjan Ghoshal

About the Author

A self researcher, writer of popular science of science education. Writes books, articles to reputed journals, magazines and news papers. Interested to think with mathematics basically in geometry, number theory, and physics to cosmology, Gravitation.

Awarded, as a fellow of mathematics form (IIK), west Bengal, INDIA.

Index

Introduction

1. A brief history of famous construction problems in Euclidean geometry
2. Operation of ruler and compass
3. Solution of angle trisection
4. Solution of cube root extraction
5. Solution of Apollonius contact problems
6. Solution of observation problems

Selected reference

THE BOOK IS DEDICATED TO MY YOUNGEST BROTHER, WHO IS NO MORE TODAY.

Introduction

Euclidean geometry known as popularly named the school geometry or classical geometry. Student learns structural science and the base of mathematics through Euclidean geometry from primary level of education.

Construction an important part of geometry and necessary for application of works in our daily life. Ignorance or unsuccessful of operations are preventing the progress of science and mathematics also the civilization of human beings.

Showing here in this book, some solutions of construction of ancient famous problems in classical geometry, these all unsolved yet. These are, angle trisection, cube root extraction, Apollonius contacts problem of circles.

This book is unique and furnishing good geometrical work of hand drawing. New methods in Euclidean geometry, I hope, these will be include to future note book of school geometry.

Using lucid English for school level students, although this book is important for all, researchers, teachers and students, although it has same importance in workers, who works with geometry.

I acknowledge with thanks to all sources of reference, and kindly direct publishing (KDP) Amazon for effortless work of publishing, distributing and promoting.

Chapter: 1

A brief history of famous construction problems in Euclidean geometry.

Euclid was not the creator of geometry. The origin of geometry was contradicted between ancient civilizations and field of works. However, the origin of geometry was very long years ago before born of Euclid. We owe to him for his valuable works, accumulation and discovered the process of study of geometry that is known as Euclidean methods or procedure or Euclidean geometry.

After published the procedure, that started development with the hands of great geometers, and problems were arisen. Three famous Greek ancient construction problems awaken, most probably nearly around 430 B C. Where 1/3rd Athenians were died off in plague. To handle the situation of political inconsistencies, great geometers although the related leaders revealed that, was a catastrophe imposes by god for lac of interest of geometry.

According to suggested era of writers of history of geometry, after then rapidly started to find the solution. Unfortunately no one can get the success really. After of long time our mathematicians are developing the concept by algebraic equation with number field consideration. They proved that, unsolved problems in Euclidean construction method.

It is known; the theory can not satisfy the result of practical. For the examples, boils law of gases and Hubble's law of expansion of universe. The both laws can calculate the zero at event of time. But the situation of volume zero impossible in matters. Therefore I can make a conjecture about construction of geometry never satisfy always completely by theoretically.

Demonstrating here the real construction for the proof of support above argument and appeal to readers for verify the construction approximately or exact.

10

Chapter: 2
Operation of ruler and compass

Our mathematical understanding of construction by ruler and compass in numbers field has been decided, only rational numbers are provided by their operation. Representing here ruler and compass's operation is not only limited to generate rational Fields only with closer property.

Introduction:

Respected mathematicians Courant and Robbins wrote in the book "what is mathematics?" Only rational operation and rational fields could be consists by ruler and compass. We know in proposition of Euclidean geometry, the even division of a given line possible by bisection with ruler and compass. (See fig I) so here p/q will be rational. Here q will be even and p will be positive integer. Therefore the construction field will be rational.

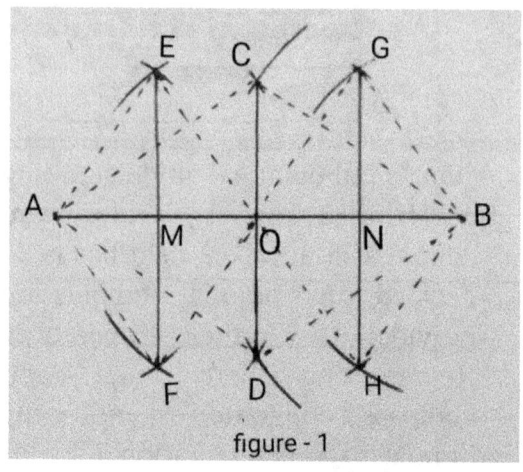
figure - 1

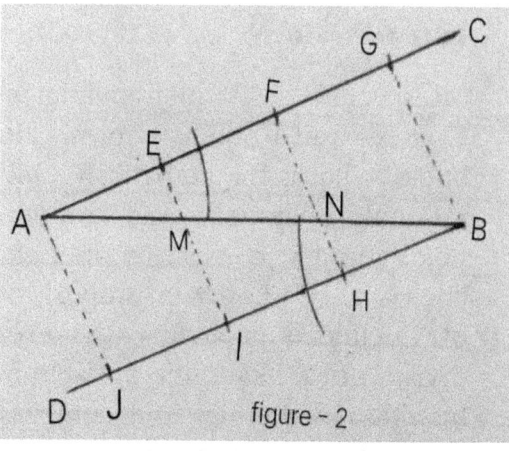
figure - 2

The operation by ruler and compass can divide the line AB to even and odd numbers, (see fig II) these operation can generate p/q rational field, and here p and q are integer. So the construction field again will have rational with closer property.

Now the construction by ruler and compass (see fig III) ACD the right angled triangle in half circle, generate AD in AB line, that the AD = $\sqrt{(AC^2 - CD^2)}$, this operation could be continue to AF and upto B, that means the operation has closer property.

But the value of AD, AF and AB will not be rational only. This operation could be generate the constructible number in extension field that may be rational and irrational both. Therefore the constructible root may be rational and irrational both. (See fig IV) Here BK the extension field have constructible numbers DF and HJ, although these are contractible root, that may be rational and irrational both.

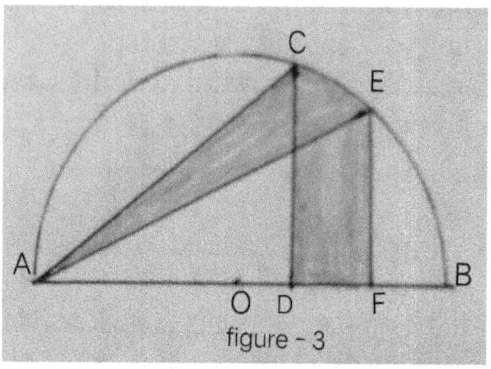
figure - 3

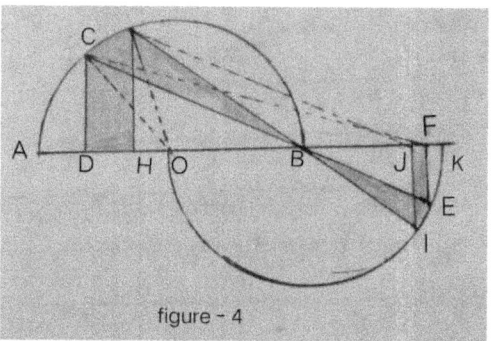
figure - 4

Suppose HJ = a+b√w, now I want to extend the operation, so I have no need in it any rational operation (addition, subtraction, multiplication, division) so DF = a + b$\sqrt{w}$ this a, b and w all are variable, that means it could be continued without any rational processes.

Conclusion:

The ruler and compass operation are not limited to only four rational processes of closer property, that argument may not be true and operation with extension field has closure property. So the value of constructible root will not be fixed and have no need of rational operation to further operation extension or contraction.

16

Chapter: 3
Solution of angle trisection

One of the three famous classical geometric construction problems is the angle trisection. Now it is accepted as impossible by ruler and compass. Is it really impossible? Showing here a construction to support of my question.

Introduction:

One of the three famous Greek classical Geometric construction problems is angle trisection. Approximate 430 B C the problem arrived. All great mathematicians were tried to solve the problem by ruler and compass, but they were unsuccessful. After that some of mathematicians provided that the impossibility by algebraic with the number field concept. Is it really impossible by ruler and compass? Is ruler and compass construction limited only on rational field? The answer below.

My contribution:

The main point of impossibility provide, compass and ruler's construction being limited to rational field and the operation can consist only rational. Here is the problem. An circle center O and diameter AB. (See fig 5) we can easily construct right angle triangle ACD and AEF, also continue the process from C to E or D to F, but here consisted field of construction from D to F are only rational numbered? Here is the base of my construction.

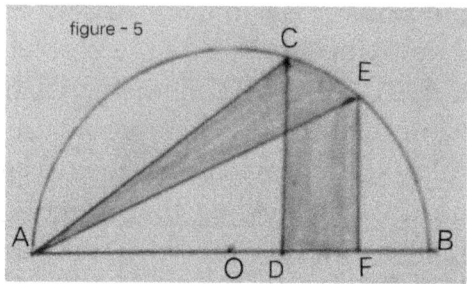

Now be the solution of construction (See fig 6). Two centers A and B of two equal radius circles, P a point between B; F, found PM = AB, and extended it up to N, so the angle NPA = 1/3rd angle of CAN, now drawn NQ through B, and join Q; P, here QP is perpendicular on BF.

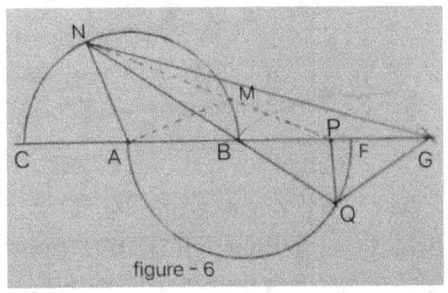

figure - 6

Suppose, QP is not perpendicular on BF, so the angle BFQ may be greater than angle GFQ, but when the P situated at F then the Q will be on arc, so the BQ and BF are not equal, this is no true. Samey if the angle GPQ be greater than angle QPB then will be arise the same problem of not equal of radius in a circle, so QP will be perpendicular exactly.

I can generalize the construction for every Angeles. (See figure – 7)

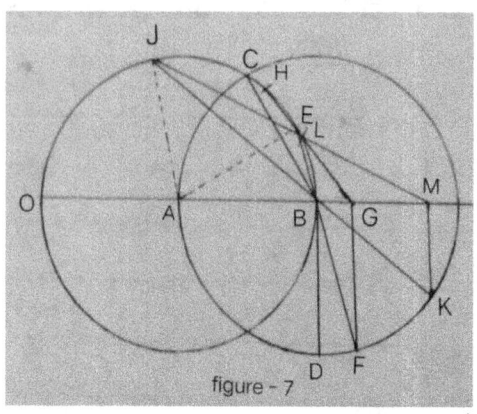

figure - 7

We know the angle ABC = 60° and DB the perpendicular on AB, it is the trisection of 180° angle OAB, another angle OAE drawing EF through B and GF perpendicular on AB, found GH = AB on arc BCO, this angle AGH = 1/3rd angle of OAE, another an angle is given OAJ draw JK through B, and perpendicular KM on AB, and found ML = AB, now the angle BML formed = 1/3rd angle of OAJ, fortunately extension of ML straight line goes through J. That was the proof of my figure – 6.

It is seen, an interesting triangle is forming through construction, those area will started from zero to zero at end. (see figure – 7) when given angle OAB = 180°, the trisecting angle is angle ABC = 60°, then formed triangle at B will be zero of area. Now be the given angle OAE, then formed triangle is EBG, and while the given angle be angle AOJ, the formed triangle will be the JBM, consequently when the given angle is angle OAO = 0°, then formed triangle will be BMB = 0°, then the M will be at end of diameter of circle B (center).

Conclusion:

I think the angle trisection problem by ruler and compass solved, so the proof of impossibility has to need further revision and consideration.

Chapter: 4
Solution of cube root extraction

Limit of operation of ruler and compass never control the development of classical geometry. Here I am showing a new theorem and development of its application can extract the cube root length of a given length of line.

Introduction:

Doubling cube, an ancient famous classical construction problem in geometry[1,2,3,4,5,6,] is a part of cube root extraction. Unsuccessfully of geometers inspired to find out the impossibility. They produced the impossibility with algebraic geometry's cubic equation and number field concept[5]. So, the problem has been accepted impossible.

Now I have find out a new theorem in classical geometry and

it's development, application give us the result of cube root length of given length of line.

Metrials and Methods:

Using classical geometric construction tools, compass and ruler and a pencil also, with obey the Euclidean method and limit of operation of ruler and compass.

My contribution:

Theorem:

If two triangles has an equal angle, the ratio of opposite sides be never change with decrease or increase of angle equally.

In triangle ABC and triangle DEF has angle ACB = angle DFE, and angle BCM = angle EFN, I have need to prove, AB/DE = AM/DN.

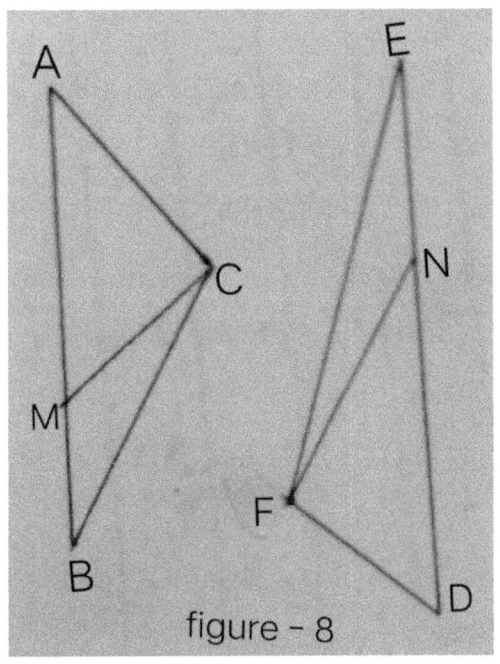

figure - 8

According to "sine rule"

$$\text{Sin ACB/AB},$$

And,

$$\text{Sin DFE/DE}$$

So, the ratio between them,

$$\text{Sin ACB/AB} : \text{sin DFE/DE}$$

Or,

$$AB/DE : \sin ACB/\sin DFE$$

Or,

$$AB/DE : 1,$$

After that,

$$\sin(ACB - BCM)/(AB - BM),$$

And,

$$\sin(DFE - EFN)/(DE - EN)$$

Therefore, the ratio,

$$\sin ACM/AM : \sin DFN/DN$$

Or,

$$AM/DN : \sin ACM/\sin DFN$$

Or,

$$AM/DN : 1$$

So,

$$AB/DE = AM/DN \text{ (proved)}$$

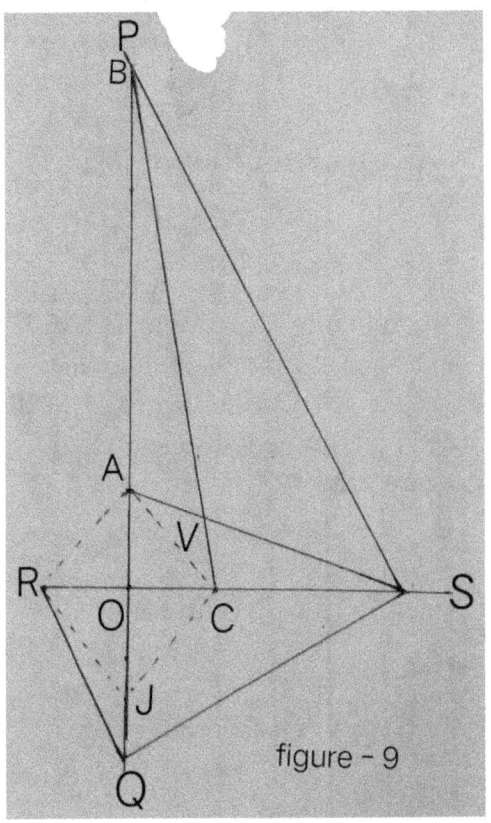

figure - 9

Now, Appling the method:

PQ and RS two straight line perpendicular situated at O, ACJN be a square of unit length or one, triangle RMD and triangle MDB are right angled, angle RMD = angle BDM = 90°, (see fig – 9)

So,

$$OB = (OM)^3,$$

And,

$$OD = (OM)^2 \text{ are formed,}$$

Here,

Sin AVB/AB : sin CVD/CD

Or,

AB/CD: sin AVB/sin CVB

Or,

AB/CD: 1,

Suppose, here OE be the length of given, (see fig – 10)

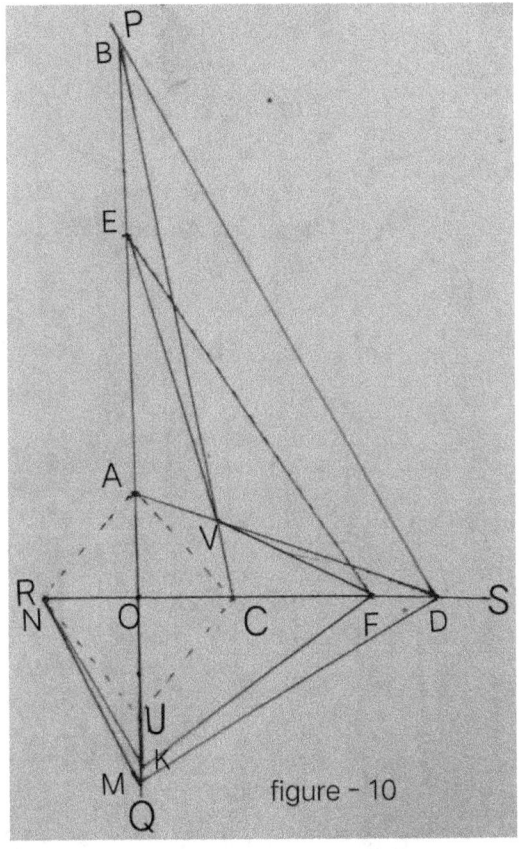

figure - 10

Join E, V and set the angle DVF = angle BE, join E, F,

So,

$$\text{Sin EVA}/EA,$$

And,

$$\text{Sin CVF}/FC$$

Or,

$$EA/FC : \sin EVA / \sin CVF$$

Or,

$$EA/FC : 1$$

So, the ratio,

$$AB/CD = EA/FC$$

Therefore,

The length $(OK)^3 = OE,$

And,

$$(OK)^2 = OF,$$

Result:

Getting cube root length OK, OE is the given length of line.

Supplementary drawing:

Showing here the solution of most significant problem the doubling cube, and cube root extraction of given length of line is eight, graphically.

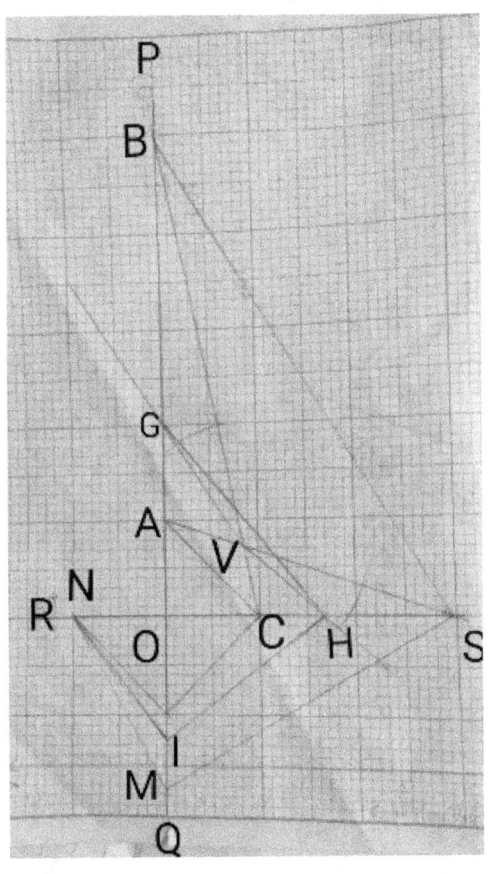

Extraction of doubling cube, here

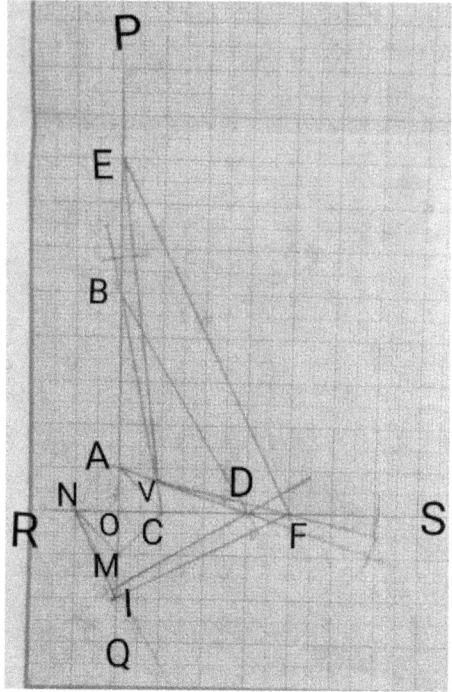

OI is the cube root of OG.

Extraction of cube root length of eight, here OI is the cube root length of OE.

Discussion:

Use of this method, cube root length extraction possible of a given length of line in classical geometry with uses of ruler and compass

Chapter: 5
Solution of Apollonius contact problems

Contact problems of circles are known as the problem of Apollonius. Famous construction problems in classical geometry by ruler and compass. Four different types of problems with three circles are unsolved till now. Representing here solutions by ruler and compass.

Introduction:

Apollonius of Perga was a great geometer as the same categories of Euclid, Archimedes and Apollonius in ancient geometric era from about 300 to 200 B.C. as the "Golden age" of Greek mathematics. He was a good astronomer. He mentioned in his work, the contact problems in three circles with different radius by

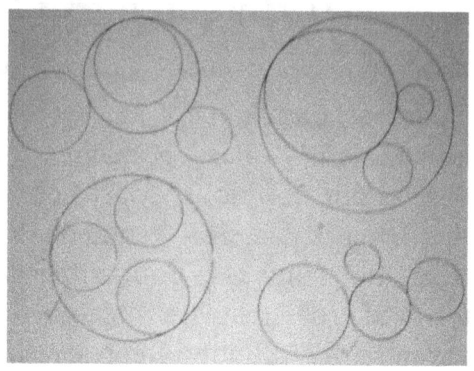

Ruler and compass. Specifically four types of problems shown in picture in book of "what is mathematics "by Courant and Robbins. Some of mathematician proved that there is possibility algebraic, but physical construction by ruler and compass has not seen yet. So the physically construction problem of Apollonius of circle unsolved yet.

My contribution:

The problems of Apollonius construction based on three variable points of three difference radius circle. Therefore the constrictions lead three basic functions. Thai is 1. Circumscribe and 2. Inscribed circle of circles. 3. Circumscribed with inscribed.

1. Circumscribed circle means, the circle will be drawn when the given circles being touch to inner wall of that circle.

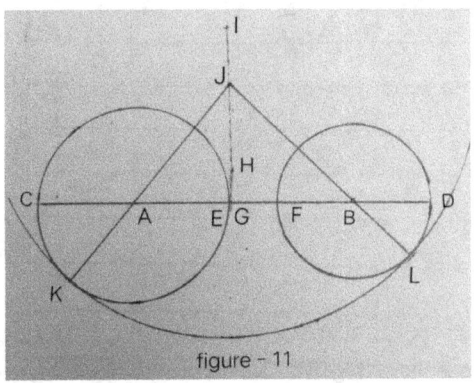
figure - 11

Here (see figure – 11) A, B center of two different radius circle. CE and DF two diameter, G mid point of CD. The H and I two points equal distance from inter wall of two circle through A, B center. Join these points by formation of arc. Now J a point on this arc, found JK and JL, this is the radius of circumscribed circle. For three circles need formation of two arcs a crossing point.

2. Inscribed circle means, the circle will be drawn when the given circles being touch to outer wall of that circle.

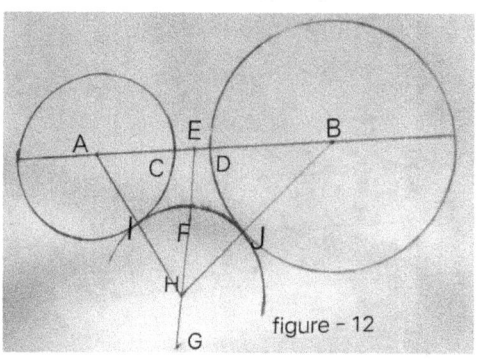

figure - 12

Here E the mid point of CDF and G two points distance from outer wall of both circles be equal, drawing an arc through E,F,G and H a point on this arc, join AH and

BH, this HI or HJ is the radius of inscribed circle. When the given circle being three then have need to formation of two arc and a cross point of them.

3. Circumscribed with inscribed means, a circle be drawing while two given circles one touch outer wall and another inner wall.(See figure – 13)

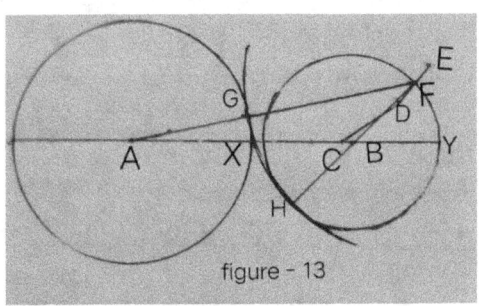

figure - 13

Here AX and BY are two radiuses of two circles. C the mid point of XY and D, E is the radius of mentioned circle, Join C, D, E points with an arc. The F is a point of that arc, FA or FH through B, this FH will be the radius of that mentioned circle.

(N: B: here using straight line than arc, for shortage of space of drawing, it may be approximate because the arc represents nearly of straight line.)

1. Construction of circumscribed circle:

A, B, C are the center of three arbitrary radius's circle. (See fig 14) drawing DE through A, B. FG through A, C and HI through C, B points. N the mid point of DE, and Q the mid point of FG. Drawing U'O = W'O and U"P = W"P, these three points joining with an arc, after that founding Q is the mid point of FG, drawing RU° = RV" and SU' = SV', now joining these points Q, R, S through with an arc, those two formed arc crossing each others at point X, these XV = XU = XW are the radius of that mentioned circle. Here XU = XV = XW, so X is the center and XU is the radius of circumscribed circle of three circles.

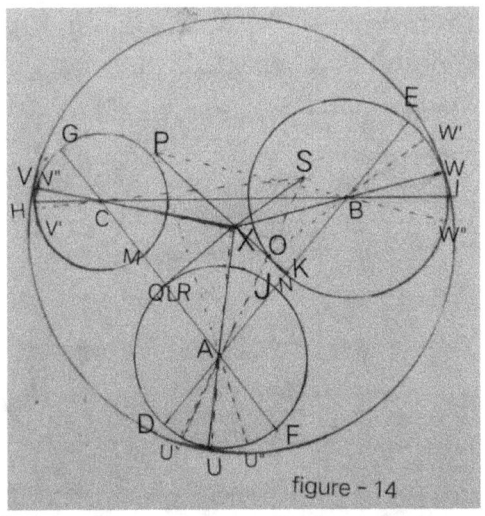

figure - 14

2. Construction of inscribed circle:

A, B, C are three center of three different radius circles, (see fig 15) N is the mid point of JK, drawing O and P those are same distance from outer wall of A and B center circles respectively. an arc is drawing through N, O, P points. Q another mid point of LM, drawing R and S are two points from same distance of outer wall of A and C radius circles respectively and drawing an arc through Q, R, S points, that formed two arcs cuts each other at X, now joining XA, XB and XC, these all lines cuts to every circles at U, V, W respectively. These length XU = XV = XW, therefore here X being the center point and XU being the radius of inscribed circle.

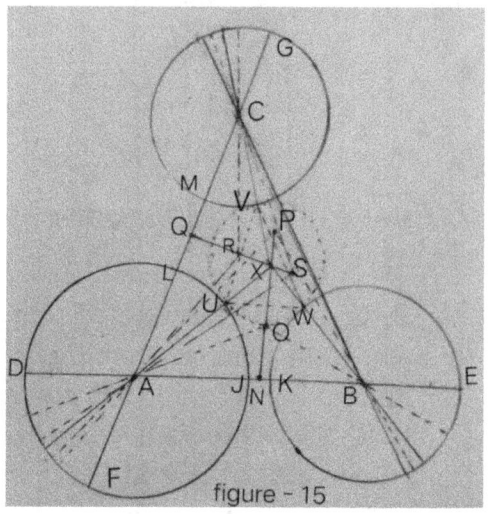

figure - 15

3. Two inscribed and one circumscribed circle's circle:

(See fig 16) H mid point of DE, drawing IU' = IW", JU" = JW", drawing an arc through points H, I,

J, now drawing M mid point of ST, and PV = PW", QV" = QW, now drawing an arc through M, P, Q, two arcs crossed at R to each others. Now the RU = RV = RW, therefore the R is center of mentioned circle and RV the radius of two subscribed and one inscribed circle's circle.

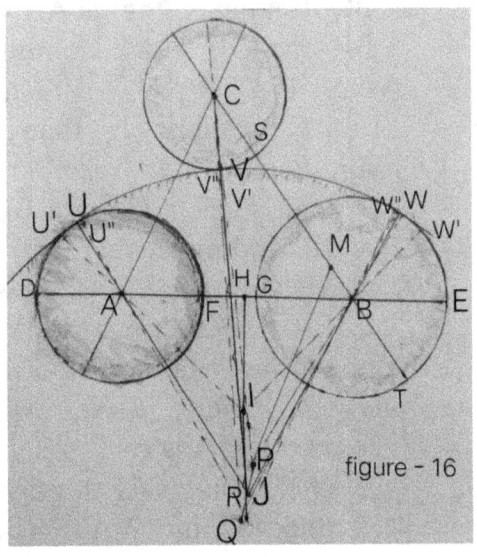

figure - 16

4. Two circumscribed and one inscribed circle's circle:

(see figure 17) A, B, C are three centers of three different radius circles. Drawing P, that is the mid point of FE, founding Q, that is AE =AQ and BF = BQ, finding R, that is AB = BR and OE = AR, joining these P, Q, R points with an arc. Now finding S, that is mid point of LM, searching T, that is CN = CT and BL = BT, finding U, that is BN = MU, drawing V, that is CM = CV and LU = BV, now drawing an arc with S, V, T points, these two formed arc cuts each others at W, drawing WX through C and joining WA, WB, these two line cuts to circle A at Z and to circle B at Y, therefore the W is the center and WX = WY = WZ radius of that mentioned circle.

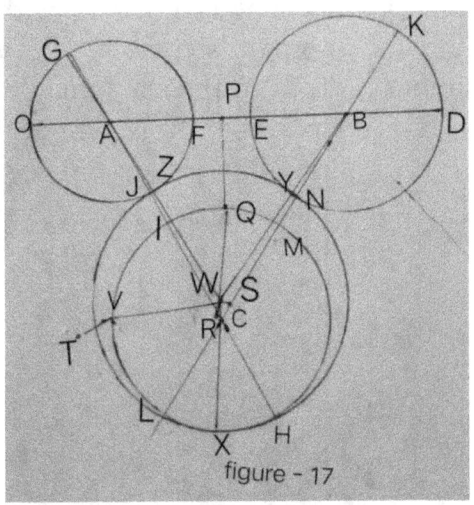

figure - 17

(N.B. using straight line except arcs, because the radius of arc being very big, so my drawing being approximate)

Key of drawing:

5. Two circumscribed circle:

A, B are two centers of different two radius circles. (See fig 18) here G the mid point of CD, finding the point H, that is BE = BH and AM

= AH, that CM = ED, finding another point I, that is BA = BI, and AN = AI, that CN = DA, joining these points I, H, G with line for approximate, now J is a point of HI, so the center J and JK = JL is the radius of formed circle.

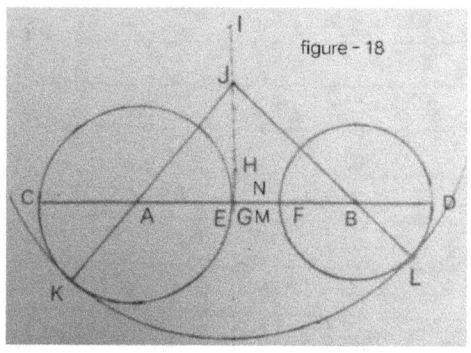

figure - 18

6. **One inscribed and one circumscribed circle:**

(See fig 19) E the mid point of AB, finding F, that is AD = AF and BC = BF, finding another point G, that is BA = BG and KD = AG, joining these points E, F, G, now H a point of GP line, joining AH and BH, therefore HI = HJ, so the center of circle is H and HI = HJ radius.

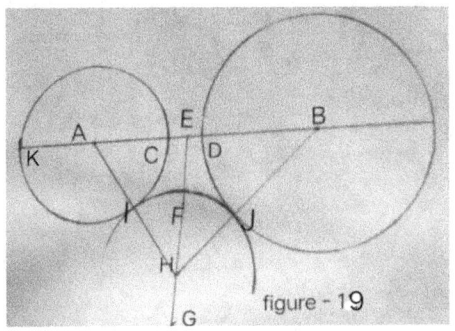

figure - 19

7. Two inscribed circle:

Here C is the mid point of XY, (see fig 20) finding the point D, this is YI = AD and IJ = BD, and the point E, that is AY = AE and BX = BE, now joining these three points C, D, E and taking F is a point of DE, joining F, A and F, H through B, therefore the center F and FG = FH is the radius of that mentioned circle.

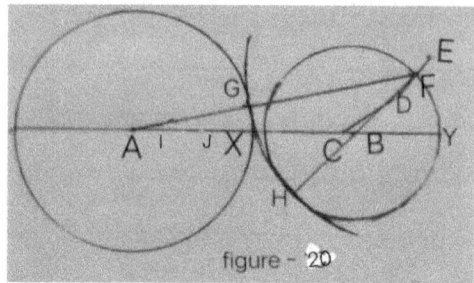

figure - 20

Conclusion:

Transformation of circle line possible to arc, but I have not seen the possibility to straight line.

Chapter: 6
Solution of observation problems

The observation of actual size and speed of a moving object understanding will not be possible without proper knowledge of classical geometry or Euclidean geometry. Discussing here the problems of observation and solution geometrically.

Introduction:

Moving or non-moving objects size and speed observation properly very difficult, and the probability of measurement will not be possible without knowledge of classical geometry. Ours eyes are the observing Metrials only. Telescope, microscope etcetera are the supporting Metrials of observation of ours eyes. The creation of image in ours retina depending on size and distance of object, but my suggestion the vision be distributed a fixed angular form. You can easily proof above suggestion.

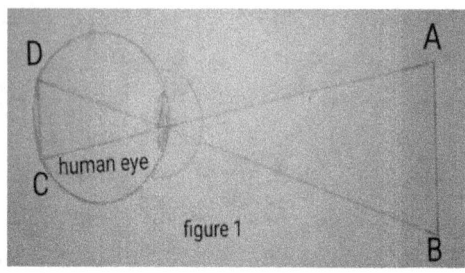

figure 1

(See figure 1) AB the object and CD is formed image in a human eye. Now you go towards AB, the object becomes large than before of you, certain a distance before you could be not seen the total object.

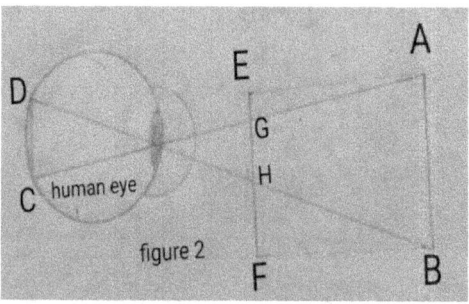

figure 2

(See figure 2)

EF is the actual size of object but you can see the size only GH. That is called visual sight. This visual sight is fixed angular form of a person of an object for fixed distance. So the clarity of image becomes too disputed while the object is going to far from him. Naturally moon's rough surface is invisible for it's far distance. Only the cause for that, our visual sight. Our visual sight gradually is large from near to far distance. So the size of an object is small than the space of visual sight. (See figure 3)

Here AB an object it is going to far CD and EF are two positions and GH is visual sight. So object size become small than visual sight's space. For that cause the speed of an object and size observation is with faulty without geometrical understanding.

(The visual sight is the visual space

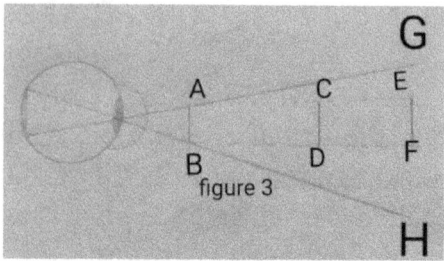

figure 3

of observation at a glance without movement of eye ball)

Size observation problem:

Stagnant object when it is situated near or far from you it's size fixed from the distance of you. So it does not create problem, but when the object is moving then it creates problem of observation of size. Specifically discussing the problems of size of planets.

We known planets in solar system is moving surrounded sun. Also our earth is moving same like them. Suppose we are the observer stagnant and others objects moving. (See fig 4)

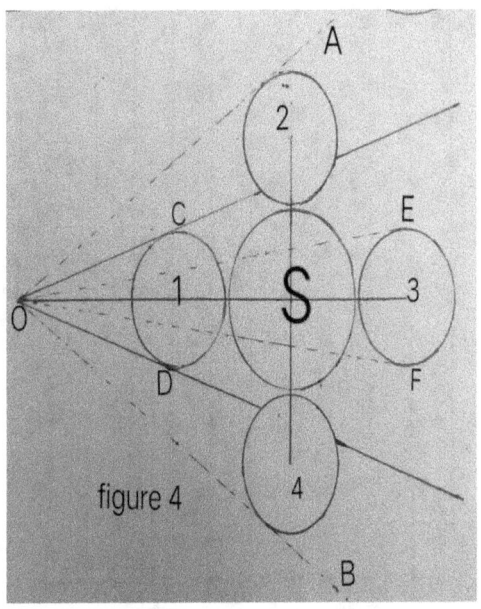

figure 4

1,2,3,4 are the position of a moving planet and S is sun, O is observer. So COD is observation angle at position 1, and EOF is observation angle at position 3, so the size calculation problem comes from it's moving States and movable position. Here geometrically exact calculation possible when we could get exact distance, I think it is very

difficult. So the determination of

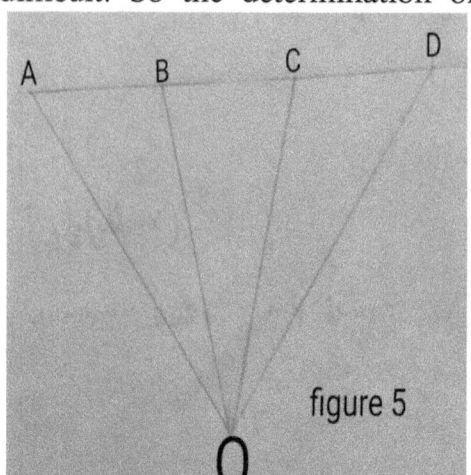

figure 5

actual size quite impossible.

Speed observation problem:

Speed observation problem is very interested; you can observe a flying airplane coming very fast towards you and going slowly to far from you. A running car comes towards you very soon but goes from you very slow. But actually the speed of airplane and car having unchanged. You understand that, what is the cause of this problem?

Here AB = BC = CD but visual sight or observation angle will not be same. The angle AOB = angle COD = 23° and angle BOC = 28°. Suppose observer's visual sight or observation angle is 28°, so the visual space will be large than AB or CD distance, then the covering area is expanded and time will be extended than before. Therefore observer seems that the speed is slowing. The coming speed rapidly shorted the distance so, observation angle becomes big so, the observer seems that the speed is increasing, but going speed rapidly doing shorted the observation angle so, observer seems that the speed is slowing.

Conclusion:

I think these problems naturally involved in observation of planetary motions. Astronomers are differentiate the speed of a planet, and the way of movement elliptical, that may be possible for the gravitation but positively have the strongly possibility of observation problem and classical geometric analysis. I think that the speed of planets never be changed and the orbit is spiral.

Selected References:

1. Carl B Boyer, *A history of mathematics,* john Wiley and sons, 1968, pg 69-176.

2. E T Bell, *The development of mathematics,* Mc-Graw Hill book company, New York, London, 1945, pg 76-79.

3. H Eves, *An introduction to the history of mathematics, 5th edition,* Sounders, New York, 1983.

4. J H Smith, *Elements of geometry,* Irvington, London, Oxford, Cambridge, 1871.

5. R Courant and H Robbins, *What is mathematics?* Oxford university press,

paperback edition, 1978, pg 120-135.

6. Sir T L Heath, *A manual of Greek mathematics,* Oxford university press, 1963, pg 154-170.

www.ingramcontent.com/pod-product-compliance
Lightning Source LLC
Chambersburg PA
CBHW070958240526
45469CB00017B/2446